Created Creatures

Student Book

Catherine McGrew Jaime

Other Books by Catherine Jaime

Illustrating History: Benjamin Franklin

Illustrating History: Leonardo da Vinci

Illustrating History: Lewis and Clark

Illustrating History: Civil Rights

Illustrating History: Slavery in the United States

A Brief History of the Lewis and Clark Expedition

The Lewis & Clark Expedition Jaime-style: "The Ultimate Field Trip"

Da Vinci: His Life and His Legacy

Leonardo the Florentine (historical fiction)

Sharing Shakespeare with Students

The Adventures of Horsey in Panama

Organized Ramblings: Home Education From A to Z

Creative Learning Connection

8006 Old Madison Pike, Ste 11-A

Madison, AL 35758

www.CreativeLearningConnection.com

Animals and Creation Science

The creatures that God has created are an amazing testimony to his creative abilities. But too many times we have let the evolutionists dictate the terms of how we study animals.

Now, with the help of groups like **Answers in Genesis** and **Vision Forum** that is changing! My family has greatly enjoyed our *Zoo Guide* and *Aquarium Guide*, our *Incredible Creatures* DVD's, and our *Jonathan Park CDs* on the *Aquarium* and the *Zoo* (and now the *Galapagos Islands*!). We have used them to help increase our knowledge of individual animals, and the amazing creative varieties in the world.

Reading is great, listening is wonderful, and watching is very helpful, but sometimes we also need to **do something**. As part of a Creation Science class I was teaching I developed the following worksheets. Each week we chose several animals to learn about. We listened to and watched portions of the above resources on those specific animals, as well as looking up information on the internet (like who their natural enemies are) and the students and I filled out the worksheets for each animal. It was a great learning experience for all of us!

I chose this number of animals in hopes of making the book comprehensive enough while keeping the book from being overwhelming to students. Don't feel compelled to have your student(s) do every worksheet, and on the other hand, if you so desire, copy the sheets and do more! One of the beauties of a program like this is its flexibility and adaptability.

I've included a "table of contents" so that students can list the animals they're studying as they go along.

Recommend Resources

Two of my favorite creation resources are **Answers in Genesis** and **Vision Forum**. We use books, dvd's, and cd's from both of these companies on a regular basis in our creation classes.

Available Through Answers in Genesis:

- *The Complete Zoo Adventure* by Gary and Mary Parker
- *The Complete Aquarium Adventure* by Merilee and Bill Clifton
- *The Incredible Creatures Volumes 1, 2, and 3* by Dr. Jobe Martin (dvd's)
- Aquarium Guide
- Zoo Guide

Available From Vision Forum:

- *Jonathan Park Goes to the Aquarium* (cd's)
- *Jonathan Park Goes to the Zoo* (cd's)
- *Jonathan Park Goes to the Galapagos* (cd's)

My Animal Pages

Name	Page #
Animal: _____	11
Animal: _____	12
Animal: _____	13
Animal: _____	14
Animal: _____	15
Animal: _____	16
Animal: _____	17
Animal: _____	18
Animal: _____	19
Animal: _____	20
Animal: _____	21
Animal: _____	22
Animal: _____	23
Animal: _____	24
Animal: _____	25
Animal: _____	26
Animal: _____	27
Animal: _____	28

Animal: _____ 29

Animal: _____ 30

Animal: _____ 31

Animal: _____ 32

Animal: _____ 33

Animal: _____ 34

Animal: _____ 35

Animal: _____ 36

Animal: _____ 37

Animal: _____ 38

Animal: _____ 39

Animal: _____ 40

Animal: _____ 41

Animal: _____ 42

Animal: _____ 43

Animal: _____ 44

Animal: _____ 45

Animal: _____ 46

Animal: _____ 47

Animal: _____ 48

Animal: _____ 49

Animal: _____ 50

Animal: _____ 51

Animal: _____ 52

Animal: _____ 53

Animal: _____ 54

Animal: _____ 55

Animal: _____ 56

Animal: _____ 57

Animal: _____ 58

Animal: _____ 59

Animal: _____ 60

Animal: _____ 61

Animal: _____ 62

Animal: _____ 63

Animal: _____ 64

Animal: _____ 65

Animal: _____ 66

Animal: _____ 67

Animal: _____ 68

Animal: _____ 69

Animal: _____ 70

Animal: _____ 71

Animal: _____ 72

Animal: _____ 73

Animal: _____ 74

Animal: _____ 75

Animal: _____ 76

Animal: _____ 77

Animal: _____ 78

Animal: _____ 79

Animal: _____ 00

Animal: _____ 81

Animal: _____ 82

Animal: _____ 83

Animal: _____ 84

Animal: _____ 85

Animal: _____ 86

Animal: _____ 87

Animal: _____ 88

Animal: _____89

Animal: _____90

Animal: _____91

Animal: _____92

Animal: _____93

Animal: _____94

Animal: _____95

Animal: _____96

Animal: _____97

Animal: _____98

Animal: _____99

Animal Name _____

Vertebrate or Invertebrate _____

Warm–blooded or Cold–blooded _____

Category (Mammal, Amphibian, Reptile, Bird, Insect, Fish, Other) _____

Size (Height/Length & Weight) _____

Habitat _____

Diet _____

Enemies _____

Created on Day _____

Draw or glue a picture of the animal above.

Specially created features _____

_____

_____

_____

Information found:
    Encyclopedia(s) _____
    On-Line _____
    Other Resources _____

Mentioned in Scripture _____

_____

Other Notes _____

_____

Animal Name _____

Vertebrate or Invertebrate _____

Warm–blooded or Cold–blooded _____

Category (Mammal, Amphibian, Reptile, Bird, Insect, Fish, Other) _____

Size (Height/Length & Weight) _____

Habitat _____

Diet _____

Enemies _____

Created on Day _____

Draw or glue a picture of the animal above.

Specially created features _____

_____

_____

_____

Information found:
    Encyclopedia(s) _____
    On-Line _____
    Other Resources _____

Mentioned in Scripture _____

_____

Other Notes _____

_____

Animal Name _____

Vertebrate or Invertebrate _____

Warm–blooded or Cold–blooded _____

Category (Mammal, Amphibian, Reptile, Bird, Insect, Fish, Other) _____

Size (Height/Length & Weight) _____

Habitat _____

Diet _____

Enemies _____

Created on Day _____

Draw or glue a picture of the animal above.

Specially created features _____

_____

_____

_____

Information found:
    Encyclopedia(s) _____
    On-Line _____
    Other Resources _____

Mentioned in Scripture _____

_____

Other Notes _____

_____

Animal Name _____

Vertebrate or Invertebrate _____

Warm–blooded or Cold–blooded _____

Category (Mammal, Amphibian, Reptile, Bird, Insect, Fish, Other) _____

Size (Height/Length & Weight) _____

Habitat _____

Diet _____

Enemies _____

Created on Day _____

Draw or glue a picture of the animal above.

Specially created features _____

_____

_____

_____

Information found:
    Encyclopedia(s) _____
    On-Line _____
    Other Resources _____

Mentioned in Scripture _____

_____

Other Notes _____

_____

Animal Name _____

Vertebrate or Invertebrate _____

Warm–blooded or Cold–blooded _____

Category (Mammal, Amphibian, Reptile, Bird, Insect, Fish, Other) _____

Size (Height/Length & Weight) _____

Habitat _____

Diet _____

Enemies _____

Created on Day _____

Draw or glue a picture of the animal above.

Specially created features _____

_____

_____

_____

Information found:
    Encyclopedia(s) _____
    On-Line _____
    Other Resources _____

Mentioned in Scripture _____

_____

Other Notes _____

_____

Animal Name _____

Vertebrate or Invertebrate _____

Warm–blooded or Cold–blooded _____

Category (Mammal, Amphibian, Reptile, Bird, Insect, Fish, Other) _____

Size (Height/Length & Weight) _____

Habitat _____

Diet _____

Enemies _____

Created on Day _____

Draw or glue a picture of the animal above.

Specially created features _____

_____

_____

_____

Information found:
    Encyclopedia(s) _____
    On-Line _____
    Other Resources _____

Mentioned in Scripture _____

_____

Other Notes _____

_____

Animal Name _____

Vertebrate or Invertebrate _____

Warm–blooded or Cold–blooded _____

Category (Mammal, Amphibian, Reptile, Bird, Insect, Fish, Other) _____

Size (Height/Length & Weight) _____

Habitat _____

Diet _____

Enemies _____

Created on Day _____

Draw or glue a picture of the animal above.

Specially created features _____

_____

_____

_____

Information found:
   Encyclopedia(s) _____
   On-Line _____
   Other Resources _____

Mentioned in Scripture _____

_____

Other Notes _____

_____

Animal Name _____

Vertebrate or Invertebrate _____

Warm–blooded or Cold–blooded _____   ·

Category (Mammal, Amphibian, Reptile, Bird, Insect, Fish, Other) _____

Size (Height/Length & Weight) _____

Habitat _____

Diet _____

Enemies _____

Created on Day _____

Draw or glue a picture of the animal above.

Specially created features _____

_____

_____

_____

Information found:
    Encyclopedia(s) _____
    On-Line _____
    Other Resources _____

Mentioned in Scripture _____

_____

Other Notes _____

_____

Animal Name _____

Vertebrate or Invertebrate _____

Warm–blooded or Cold–blooded _____

Category (Mammal, Amphibian, Reptile, Bird, Insect, Fish, Other) _____

Size (Height/Length & Weight) _____

Habitat _____

Diet _____

Enemies _____

Created on Day _____

Draw or glue a picture of the animal above.

Specially created features _____

_____

_____

_____

Information found:
  Encyclopedia(s) _____
  On-Line _____
  Other Resources _____

Mentioned in Scripture _____

_____

Other Notes _____

_____

Animal Name _____

Vertebrate or Invertebrate _____

Warm–blooded or Cold–blooded _____

Category (Mammal, Amphibian, Reptile, Bird, Insect, Fish, Other) _____

Size (Height/Length & Weight) _____

Habitat _____

Diet _____

Enemies _____

Created on Day _____

Draw or glue a picture of the animal above.

Specially created features _____

_____

_____

_____

Information found:
    Encyclopedia(s) _____
    On-Line _____
    Other Resources _____

Mentioned in Scripture _____

_____

Other Notes _____

_____

Animal Name _____

Vertebrate or Invertebrate _____

Warm–blooded or Cold–blooded _____

Category (Mammal, Amphibian, Reptile, Bird, Insect, Fish, Other) _____

Size (Height/Length & Weight) _____

Habitat _____

Diet _____

Enemies _____

Created on Day _____

Draw or glue a picture of the animal above.

Specially created features _____

_____

_____

_____

Information found:
    Encyclopedia(s) _____

    On-Line _____

    Other Resources _____

Mentioned in Scripture _____

_____

Other Notes _____

_____

Animal Name _____

Vertebrate or Invertebrate _____

Warm–blooded or Cold–blooded _____

Category (Mammal, Amphibian, Reptile, Bird, Insect, Fish, Other) _____

Size (Height/Length & Weight) _____

Habitat _____

Diet _____

Enemies _____

Created on Day _____

Draw or glue a picture of the animal above.

Specially created features _____

_____

_____

_____

Information found:
    Encyclopedia(s) _____
    On-Line _____
    Other Resources _____

Mentioned in Scripture _____

_____

Other Notes _____

_____

Animal Name _____

Vertebrate or Invertebrate _____

Warm–blooded or Cold–blooded _____

Category (Mammal, Amphibian, Reptile, Bird, Insect, Fish, Other) _____

Size (Height/Length & Weight) _____

Habitat _____

Diet _____

Enemies _____

Created on Day _____

Draw or glue a picture of the animal above.

Specially created features _____

_____

_____

_____

Information found:
    Encyclopedia(s) _____
    On-Line _____
    Other Resources _____

Mentioned in Scripture _____

_____

Other Notes _____

_____

Animal Name _____

Vertebrate or Invertebrate _____

Warm–blooded or Cold–blooded _____

Category (Mammal, Amphibian, Reptile, Bird, Insect, Fish, Other) _____

Size (Height/Length & Weight) _____

Habitat _____

Diet _____

Enemies _____

Created on Day _____

Draw or glue a picture of the animal above.

Specially created features _____

_____

_____

_____

Information found:
    Encyclopedia(s) _____
    On-Line _____
    Other Resources _____

Mentioned in Scripture _____

_____

Other Notes _____

_____

Animal Name _____

Vertebrate or Invertebrate _____

Warm–blooded or Cold–blooded _____

Category (Mammal, Amphibian, Reptile, Bird, Insect, Fish, Other) _____

Size (Height/Length & Weight) _____

Habitat _____

Diet _____

Enemies _____

Created on Day _____

Draw or glue a picture of the animal above.

Specially created features _____

_____

_____

_____

Information found:
    Encyclopedia(s) _____
    On-Line _____
    Other Resources _____

Mentioned in Scripture _____

_____

Other Notes _____

_____

Animal Name _____

Vertebrate or Invertebrate _____

Warm–blooded or Cold–blooded _____

Category (Mammal, Amphibian, Reptile, Bird, Insect, Fish, Other) _____

Size (Height/Length & Weight) _____

Habitat _____

Diet _____

Enemies _____

Created on Day _____

Draw or glue a picture of the animal above.

Specially created features _____

_____

_____

_____

Information found:
    Encyclopedia(s) _____
    On-Line _____
    Other Resources _____

Mentioned in Scripture _____

_____

Other Notes _____

_____

Animal Name _____

Vertebrate or Invertebrate _____

Warm–blooded or Cold–blooded _____

Category (Mammal, Amphibian, Reptile, Bird, Insect, Fish, Other) _____

Size (Height/Length & Weight) _____

Habitat _____

Diet _____

Enemies _____

Created on Day _____

Draw or glue a picture of the animal above.

Specially created features _____

_____

_____

_____

Information found:
    Encyclopedia(s) _____
    On-Line _____
    Other Resources _____

Mentioned in Scripture _____

_____

Other Notes _____

_____

Animal Name _____

Vertebrate or Invertebrate _____

Warm–blooded or Cold–blooded _____

Category (Mammal, Amphibian, Reptile, Bird, Insect, Fish, Other) _____

Size (Height/Length & Weight) _____

Habitat _____

Diet _____

Enemies _____

Created on Day _____

Draw or glue a picture of the animal above.

Specially created features _____

_____

_____

_____

Information found:
    Encyclopedia(s) _____
    On-Line _____
    Other Resources _____

Mentioned in Scripture _____

_____

Other Notes _____

_____

Animal Name _____

Vertebrate or Invertebrate _____

Warm–blooded or Cold–blooded _____

Category (Mammal, Amphibian, Reptile, Bird, Insect, Fish, Other) _____

Size (Height/Length & Weight) _____

Habitat _____

Diet _____

Enemies _____

Created on Day _____

Draw or glue a picture of the animal above.

Specially created features _____

_____

_____

_____

Information found:
    Encyclopedia(s) _____
    On-Line _____
    Other Resources _____

Mentioned in Scripture _____

_____

Other Notes _____

_____

Animal Name _____

Vertebrate or Invertebrate _____

Warm–blooded or Cold–blooded _____

Category (Mammal, Amphibian, Reptile, Bird, Insect, Fish, Other) _____

Size (Height/Length & Weight) _____

Habitat _____

Diet _____

Enemies _____

Created on Day _____

Draw or glue a picture of the animal above.

Specially created features _____

_____

_____

_____

Information found:
    Encyclopedia(s) _____

    On-Line _____

    Other Resources _____

Mentioned in Scripture _____

_____

Other Notes _____

_____

Animal Name _____

Vertebrate or Invertebrate _____

Warm–blooded or Cold–blooded _____

Category (Mammal, Amphibian, Reptile, Bird, Insect, Fish, Other) _____

Size (Height/Length & Weight) _____

Habitat _____

Diet _____

Enemies _____

Created on Day _____

Draw or glue a picture of the animal above.

Specially created features _____

_____

_____

_____

Information found:
    Encyclopedia(s) _____
    On-Line _____
    Other Resources _____

Mentioned in Scripture _____

_____

Other Notes _____

_____

Animal Name _____

Vertebrate or Invertebrate _____

Warm–blooded or Cold–blooded _____

Category (Mammal, Amphibian, Reptile, Bird, Insect, Fish, Other) _____

Size (Height/Length & Weight) _____

Habitat _____

Diet _____

Enemies _____

Created on Day _____

Draw or glue a picture of the animal above.

Specially created features _____

_____

_____

_____

Information found:
    Encyclopedia(s) _____
    On-Line _____
    Other Resources _____

Mentioned in Scripture _____

_____

Other Notes _____

_____

Animal Name _____

Vertebrate or Invertebrate _____

Warm–blooded or Cold–blooded _____

Category (Mammal, Amphibian, Reptile, Bird, Insect, Fish, Other) _____

Size (Height/Length & Weight) _____

Habitat _____

Diet _____

Enemies _____

Created on Day _____

Draw or glue a picture of the animal above.

Specially created features _____

_____

_____

_____

Information found:
    Encyclopedia(s) _____
    On-Line _____
    Other Resources _____

Mentioned in Scripture _____

_____

Other Notes _____

_____

Animal Name _____

Vertebrate or Invertebrate _____

Warm–blooded or Cold–blooded _____

Category (Mammal, Amphibian, Reptile, Bird, Insect, Fish, Other) _____

Size (Height/Length & Weight) _____

Habitat _____

Diet _____

Enemies _____

Created on Day _____

Draw or glue a picture of the animal above.

Specially created features _____

_____

_____

_____

Information found:
    Encyclopedia(s) _____
    On-Line _____
    Other Resources _____

Mentioned in Scripture _____

_____

Other Notes _____

_____

Animal Name _____

Vertebrate or Invertebrate _____

Warm–blooded or Cold–blooded _____

Category (Mammal, Amphibian, Reptile, Bird, Insect, Fish, Other) _____

Size (Height/Length & Weight) _____

Habitat _____

Diet _____

Enemies _____

Created on Day _____

Draw or glue a picture of the animal above.

Specially created features _____

_____

_____

_____

Information found:
    Encyclopedia(s) _____
    On-Line _____
    Other Resources _____

Mentioned in Scripture _____

_____

Other Notes _____

_____

Animal Name _____

Vertebrate or Invertebrate _____

Warm–blooded or Cold–blooded _____

Category (Mammal, Amphibian, Reptile, Bird, Insect, Fish, Other) _____

Size (Height/Length & Weight) _____

Habitat _____

Diet _____

Enemies _____

Created on Day _____

Draw or glue a picture of the animal above.

Specially created features _____

_____

_____

_____

Information found:
    Encyclopedia(s) _____

    On-Line _____

    Other Resources _____

Mentioned in Scripture _____

_____

Other Notes _____

_____

Animal Name _____

Vertebrate or Invertebrate _____

Warm–blooded or Cold–blooded _____

Category (Mammal, Amphibian, Reptile, Bird, Insect, Fish, Other) _____

Size (Height/Length & Weight) _____

Habitat _____

Diet _____

Enemies _____

Created on Day _____

Draw or glue a picture of the animal above.

Specially created features _____

_____

_____

_____

Information found:
    Encyclopedia(s) _____
    On-Line _____
    Other Resources _____

Mentioned in Scripture _____

_____

Other Notes _____

_____

Animal Name _____

Vertebrate or Invertebrate _____

Warm–blooded or Cold–blooded _____

Category (Mammal, Amphibian, Reptile, Bird, Insect, Fish, Other) _____

Size (Height/Length & Weight) _____

Habitat _____

Diet _____

Enemies _____

Created on Day _____

Draw or glue a picture of the animal above.

Specially created features _____

_____

_____

_____

Information found:
    Encyclopedia(s) _____
    On-Line _____
    Other Resources _____

Mentioned in Scripture _____

_____

Other Notes _____

_____

Animal Name _____

Vertebrate or Invertebrate _____

Warm–blooded or Cold–blooded _____

Category (Mammal, Amphibian, Reptile, Bird, Insect, Fish, Other) _____

Size (Height/Length & Weight) _____

Habitat _____

Diet _____

Enemies _____

Created on Day _____

Draw or glue a picture of the animal above.

Specially created features _____

_____

_____

_____

Information found:
    Encyclopedia(s) _____
    On-Line _____
    Other Resources _____

Mentioned in Scripture _____

_____

Other Notes _____

_____

Animal Name _____

Vertebrate or Invertebrate _____

Warm–blooded or Cold–blooded _____

Category (Mammal, Amphibian, Reptile, Bird, Insect, Fish, Other) _____

Size (Height/Length & Weight) _____

Habitat _____

Diet _____

Enemies _____

Created on Day _____

Draw or glue a picture of the animal above.

Specially created features _____

_____

_____

_____

Information found:
    Encyclopedia(s) _____
    On-Line _____
    Other Resources _____

Mentioned in Scripture _____

_____

Other Notes _____

_____

Animal Name _____

Vertebrate or Invertebrate _____

Warm–blooded or Cold–blooded _____

Category (Mammal, Amphibian, Reptile, Bird, Insect, Fish, Other) _____

Size (Height/Length & Weight) _____

Habitat _____

Diet _____

Enemies _____

Created on Day _____

Draw or glue a picture of the animal above.

Specially created features _____

_____

_____

_____

Information found:
    Encyclopedia(s) _____
    On-Line _____
    Other Resources _____

Mentioned in Scripture _____

_____

Other Notes _____

_____

Animal Name _____

Vertebrate or Invertebrate _____

Warm–blooded or Cold–blooded _____

Category (Mammal, Amphibian, Reptile, Bird, Insect, Fish, Other) _____

Size (Height/Length & Weight) _____

Habitat _____

Diet _____

Enemies _____

Created on Day _____

Draw or glue a picture of the animal above.

Specially created features _____

_____

_____

_____

Information found:
    Encyclopedia(s) _____

    On-Line _____

    Other Resources _____

Mentioned in Scripture _____

_____

Other Notes _____

_____

Animal Name _____

Vertebrate or Invertebrate _____

Warm–blooded or Cold–blooded _____

Category (Mammal, Amphibian, Reptile, Bird, Insect, Fish, Other) _____

Size (Height/Length & Weight) _____

Habitat _____

Diet _____

Enemies _____

Created on Day _____

Draw or glue a picture of the animal above.

Specially created features _____

_____

_____

_____

Information found:
    Encyclopedia(s) _____
    On-Line _____
    Other Resources _____

Mentioned in Scripture _____

_____

Other Notes _____

_____

Animal Name _____

Vertebrate or Invertebrate _____

Warm–blooded or Cold–blooded _____

Category (Mammal, Amphibian, Reptile, Bird, Insect, Fish, Other) _____

Size (Height/Length & Weight) _____

Habitat _____

Diet _____

Enemies _____

Created on Day _____

Draw or glue a picture of the animal above.

Specially created features _____

_____

_____

_____

Information found:

    Encyclopedia(s) _____

    On-Line _____

    Other Resources _____

Mentioned in Scripture _____

_____

Other Notes _____

_____

Animal Name _____

Vertebrate or Invertebrate _____

Warm–blooded or Cold–blooded _____

Category (Mammal, Amphibian, Reptile, Bird, Insect, Fish, Other) _____

Size (Height/Length & Weight) _____

Habitat _____

Diet _____

Enemies _____

Created on Day _____

Draw or glue a picture of the animal above.

Specially created features _____

_____

_____

_____

Information found:
    Encyclopedia(s) _____
    On-Line _____
    Other Resources _____

Mentioned in Scripture _____

_____

Other Notes _____

_____

Animal Name _____

Vertebrate or Invertebrate _____

Warm–blooded or Cold–blooded _____

Category (Mammal, Amphibian, Reptile, Bird, Insect, Fish, Other) _____

Size (Height/Length & Weight) _____

Habitat _____

Diet _____

Enemies _____

Created on Day _____

Draw or glue a picture of the animal above.

Specially created features _____

_____

_____

_____

Information found:
    Encyclopedia(s) _____
    On-Line _____
    Other Resources _____

Mentioned in Scripture _____

_____

Other Notes _____

_____

Animal Name _____

Vertebrate or Invertebrate _____

Warm–blooded or Cold–blooded _____

Category (Mammal, Amphibian, Reptile, Bird, Insect, Fish, Other) _____

Size (Height/Length & Weight) _____

Habitat _____

Diet _____

Enemies _____

Created on Day _____

Draw or glue a picture of the animal above.

Specially created features _____

_____

_____

_____

Information found:
    Encyclopedia(s) _____
    On-Line _____
    Other Resources _____

Mentioned in Scripture _____

_____

Other Notes _____

_____

Animal Name _____

Vertebrate or Invertebrate _____

Warm–blooded or Cold–blooded _____

Category (Mammal, Amphibian, Reptile, Bird, Insect, Fish, Other) _____

Size (Height/Length & Weight) _____

Habitat _____

Diet _____

Enemies _____

Created on Day _____

Draw or glue a picture of the animal above.

Specially created features _____

_____

_____

_____

Information found:
    Encyclopedia(s) _____
    On-Line _____
    Other Resources _____

Mentioned in Scripture _____

_____

Other Notes _____

_____

Animal Name _____

Vertebrate or Invertebrate _____

Warm–blooded or Cold–blooded _____

Category (Mammal, Amphibian, Reptile, Bird, Insect, Fish, Other) _____

Size (Height/Length & Weight) _____

Habitat _____

Diet _____

Enemies _____

Created on Day _____

Draw or glue a picture of the animal above.

Specially created features _____

_____

_____

_____

Information found:
    Encyclopedia(s) _____
    On-Line _____
    Other Resources _____

Mentioned in Scripture _____

_____

Other Notes _____

_____

Animal Name _____

Vertebrate or Invertebrate _____

Warm–blooded or Cold–blooded _____

Category (Mammal, Amphibian, Reptile, Bird, Insect, Fish, Other) _____

Size (Height/Length & Weight) _____

Habitat _____

Diet _____

Enemies _____

Created on Day _____

Draw or glue a picture of the animal above.

Specially created features _____

_____

_____

_____

Information found:
    Encyclopedia(s) _____
    On-Line _____
    Other Resources _____

Mentioned in Scripture _____

_____

Other Notes _____

_____

Animal Name _____

Vertebrate or Invertebrate _____

Warm–blooded or Cold–blooded _____

Category (Mammal, Amphibian, Reptile, Bird, Insect, Fish, Other) _____

Size (Height/Length & Weight) _____

Habitat _____

Diet _____

Enemies _____

Created on Day _____

Draw or glue a picture of the animal above.

Specially created features _____

_____

_____

_____

Information found:
   Encyclopedia(s) _____
   On-Line _____
   Other Resources _____

Mentioned in Scripture _____

_____

Other Notes _____

_____

Animal Name _____

Vertebrate or Invertebrate _____

Warm–blooded or Cold–blooded _____

Category (Mammal, Amphibian, Reptile, Bird, Insect, Fish, Other) _____

Size (Height/Length & Weight) _____

Habitat _____

Diet _____

Enemies _____

Created on Day _____

Draw or glue a picture of the animal above.

Specially created features _____

_____

_____

_____

Information found:
     Encyclopedia(s) _____
     On-Line _____
     Other Resources _____

Mentioned in Scripture _____

_____

Other Notes _____

_____

Animal Name _____

Vertebrate or Invertebrate _____

Warm–blooded or Cold–blooded _____

Category (Mammal, Amphibian, Reptile, Bird, Insect, Fish, Other) _____

Size (Height/Length & Weight) _____

Habitat _____

Diet _____

Enemies _____

Created on Day _____

Draw or glue a picture of the animal above.

Specially created features _____

_____

_____

_____

Information found:
    Encyclopedia(s) _____
    On-Line _____
    Other Resources _____

Mentioned in Scripture _____

_____

Other Notes _____

_____

Animal Name _____

Vertebrate or Invertebrate _____

Warm–blooded or Cold–blooded _____

Category (Mammal, Amphibian, Reptile, Bird, Insect, Fish, Other) _____

Size (Height/Length & Weight) _____

Habitat _____

Diet _____

Enemies _____

Created on Day _____

Draw or glue a picture of the animal above.

Specially created features _____

_____

_____

_____

Information found:
    Encyclopedia(s) _____
    On-Line _____
    Other Resources _____

Mentioned in Scripture _____

_____

Other Notes _____

_____

Animal Name _____

Vertebrate or Invertebrate _____

Warm–blooded or Cold–blooded _____

Category (Mammal, Amphibian, Reptile, Bird, Insect, Fish, Other) _____

Size (Height/Length & Weight) _____

Habitat _____

Diet _____

Enemies _____

Created on Day _____

Draw or glue a picture of the animal above.

Specially created features _____

_____

_____

_____

Information found:
    Encyclopedia(s) _____
    On-Line _____
    Other Resources _____

Mentioned in Scripture _____

_____

Other Notes _____

_____

Animal Name _____

Vertebrate or Invertebrate _____

Warm–blooded or Cold–blooded _____

Category (Mammal, Amphibian, Reptile, Bird, Insect, Fish, Other) _____

Size (Height/Length & Weight) _____

Habitat _____

Diet _____

Enemies _____

Created on Day _____

Draw or glue a picture of the animal above.

Specially created features _____

_____

_____

_____

Information found:
    Encyclopedia(s) _____
    On-Line _____
    Other Resources _____

Mentioned in Scripture _____

_____

Other Notes _____

_____

Animal Name _____

Vertebrate or Invertebrate _____

Warm–blooded or Cold–blooded _____

Category (Mammal, Amphibian, Reptile, Bird, Insect, Fish, Other) _____

Size (Height/Length & Weight) _____

Habitat _____

Diet _____

Enemies _____

Created on Day _____

Draw or glue a picture of the animal above.

Specially created features _____

_____

_____

_____

Information found:
    Encyclopedia(s) _____
    On-Line _____
    Other Resources _____

Mentioned in Scripture _____

_____

Other Notes _____

_____

Animal Name _____

Vertebrate or Invertebrate _____

Warm–blooded or Cold–blooded _____

Category (Mammal, Amphibian, Reptile, Bird, Insect, Fish, Other) _____

Size (Height/Length & Weight) _____

Habitat _____

Diet _____

Enemies _____

Created on Day _____

Draw or glue a picture of the animal above.

Specially created features _____

_____

_____

_____

Information found:
     Encyclopedia(s) _____
     On-Line _____
     Other Resources _____

Mentioned in Scripture _____

_____

Other Notes _____

_____

Animal Name _____

Vertebrate or Invertebrate _____

Warm–blooded or Cold–blooded _____

Category (Mammal, Amphibian, Reptile, Bird, Insect, Fish, Other) _____

Size (Height/Length & Weight) _____

Habitat _____

Diet _____

Enemies _____

Created on Day _____

Draw or glue a picture of the animal above.

Specially created features _____

_____

_____

_____

Information found:
    Encyclopedia(s) _____
    On-Line _____
    Other Resources _____

Mentioned in Scripture _____

_____

Other Notes _____

_____

Animal Name _____

Vertebrate or Invertebrate _____

Warm–blooded or Cold–blooded _____

Category (Mammal, Amphibian, Reptile, Bird, Insect, Fish, Other) _____

Size (Height/Length & Weight) _____

Habitat _____

Diet _____

Enemies _____

Created on Day _____

Draw or glue a picture of the animal above.

Specially created features _____

_____

_____

_____

Information found:
    Encyclopedia(s) _____
    On-Line _____
    Other Resources _____

Mentioned in Scripture _____

_____

Other Notes _____

_____

Animal Name _____

Vertebrate or Invertebrate _____

Warm–blooded or Cold–blooded _____

Category (Mammal, Amphibian, Reptile, Bird, Insect, Fish, Other) _____

Size (Height/Length & Weight) _____

Habitat _____

Diet _____

Enemies _____

Created on Day _____

Draw or glue a picture of the animal above.

Specially created features _____

_____

_____

_____

Information found:
    Encyclopedia(s) _____
    On-Line _____
    Other Resources _____

Mentioned in Scripture _____

_____

Other Notes _____

_____

Animal Name _____

Vertebrate or Invertebrate _____

Warm–blooded or Cold–blooded _____

Category (Mammal, Amphibian, Reptile, Bird, Insect, Fish, Other) _____

Size (Height/Length & Weight) _____

Habitat _____

Diet _____

Enemies _____

Created on Day _____

Draw or glue a picture of the animal above.

Specially created features _____

_____

_____

_____

Information found:
    Encyclopedia(s) _____
    On-Line _____
    Other Resources _____

Mentioned in Scripture _____

_____

Other Notes _____

_____

Animal Name _____

Vertebrate or Invertebrate _____

Warm–blooded or Cold–blooded _____

Category (Mammal, Amphibian, Reptile, Bird, Insect, Fish, Other) _____

Size (Height/Length & Weight) _____

Habitat _____

Diet _____

Enemies _____

Created on Day _____

Draw or glue a picture of the animal above.

Specially created features _____

_____

_____

_____

Information found:
    Encyclopedia(s) _____
    On-Line _____
    Other Resources _____

Mentioned in Scripture _____

_____

Other Notes _____

_____

Animal Name _____

Vertebrate or Invertebrate _____

Warm–blooded or Cold–blooded _____

Category (Mammal, Amphibian, Reptile, Bird, Insect, Fish, Other) _____

Size (Height/Length & Weight) _____

Habitat _____

Diet _____

Enemies _____

Created on Day _____

Draw or glue a picture of the animal above.

Specially created features _____

_____

_____

_____

Information found:
    Encyclopedia(s) _____
    On-Line _____
    Other Resources _____

Mentioned in Scripture _____

_____

Other Notes _____

_____

Animal Name _____

Vertebrate or Invertebrate _____

Warm–blooded or Cold–blooded _____

Category (Mammal, Amphibian, Reptile, Bird, Insect, Fish, Other) _____

Size (Height/Length & Weight) _____

Habitat _____

Diet _____

Enemies _____

Created on Day _____

Draw or glue a picture of the animal above.

Specially created features _____

_____

_____

_____

Information found:
    Encyclopedia(s) _____
    On-Line _____
    Other Resources _____

Mentioned in Scripture _____

_____

Other Notes _____

_____

Animal Name _____

Vertebrate or Invertebrate _____

Warm–blooded or Cold–blooded _____

Category (Mammal, Amphibian, Reptile, Bird, Insect, Fish, Other) _____

Size (Height/Length & Weight) _____

Habitat _____

Diet _____

Enemies _____

Created on Day _____

Draw or glue a picture of the animal above.

Specially created features _____

_____

_____

_____

Information found:
    Encyclopedia(s) _____
    On-Line _____
    Other Resources _____

Mentioned in Scripture _____

_____

Other Notes _____

_____

Animal Name _____

Vertebrate or Invertebrate _____

Warm–blooded or Cold–blooded _____

Category (Mammal, Amphibian, Reptile, Bird, Insect, Fish, Other) _____

Size (Height/Length & Weight) _____

Habitat _____

Diet _____

Enemies _____

Created on Day _____

Draw or glue a picture of the animal above.

Specially created features _____

_____

_____

_____

Information found:
    Encyclopedia(s) _____

    On-Line _____

    Other Resources _____

Mentioned in Scripture _____

_____

Other Notes _____

_____

Animal Name _____

Vertebrate or Invertebrate _____

Warm–blooded or Cold–blooded _____

Category (Mammal, Amphibian, Reptile, Bird, Insect, Fish, Other) _____

Size (Height/Length & Weight) _____

Habitat _____

Diet _____

Enemies _____

Created on Day _____

Draw or glue a picture of the animal above.

Specially created features _____

_____

_____

_____

Information found:
    Encyclopedia(s) _____
    On-Line _____
    Other Resources _____

Mentioned in Scripture _____

_____

Other Notes _____

_____

Animal Name _____

Vertebrate or Invertebrate _____

Warm–blooded or Cold–blooded _____

Category (Mammal, Amphibian, Reptile, Bird, Insect, Fish, Other) _____

Size (Height/Length & Weight) _____

Habitat _____

Diet _____

Enemies _____

Created on Day _____

Draw or glue a picture of the animal above.

Specially created features _____

_____

_____

_____

Information found:
    Encyclopedia(s) _____
    On-Line _____
    Other Resources _____

Mentioned in Scripture _____

_____

Other Notes _____

_____

Animal Name _____

Vertebrate or Invertebrate _____

Warm–blooded or Cold–blooded _____

Category (Mammal, Amphibian, Reptile, Bird, Insect, Fish, Other) _____

Size (Height/Length & Weight) _____

Habitat _____

Diet _____

Enemies _____

Created on Day _____

Draw or glue a picture of the animal above.

Specially created features _____

_____

_____

_____

Information found:
    Encyclopedia(s) _____
    On-Line _____
    Other Resources _____

Mentioned in Scripture _____

_____

Other Notes _____

_____

Animal Name _____

Vertebrate or Invertebrate _____

Warm–blooded or Cold–blooded _____

Category (Mammal, Amphibian, Reptile, Bird, Insect, Fish, Other) _____

Size (Height/Length & Weight) _____

Habitat _____

Diet _____

Enemies _____

Created on Day _____

Draw or glue a picture of the animal above.

Specially created features _____

_____

_____

_____

Information found:
    Encyclopedia(s) _____
    On-Line _____
    Other Resources _____

Mentioned in Scripture _____

_____

Other Notes _____

_____

Animal Name _____

Vertebrate or Invertebrate _____

Warm–blooded or Cold–blooded _____

Category (Mammal, Amphibian, Reptile, Bird, Insect, Fish, Other) _____

Size (Height/Length & Weight) _____

Habitat _____

Diet _____

Enemies _____

Created on Day _____

Draw or glue a picture of the animal above.

Specially created features _____

_____

_____

_____

Information found:
    Encyclopedia(s) _____
    On-Line _____
    Other Resources _____

Mentioned in Scripture _____

_____

Other Notes _____

_____

Animal Name _____

Vertebrate or Invertebrate _____

Warm–blooded or Cold–blooded _____

Category (Mammal, Amphibian, Reptile, Bird, Insect, Fish, Other) _____

Size (Height/Length & Weight) _____

Habitat _____

Diet _____

Enemies _____

Created on Day _____

Draw or glue a picture of the animal above.

Specially created features _____

_____

_____

_____

Information found:
    Encyclopedia(s) _____
    On-Line _____
    Other Resources _____

Mentioned in Scripture _____

_____

Other Notes _____

_____

Animal Name _____

Vertebrate or Invertebrate _____

Warm–blooded or Cold–blooded _____

Category (Mammal, Amphibian, Reptile, Bird, Insect, Fish, Other) _____

Size (Height/Length & Weight) _____

Habitat _____

Diet _____

Enemies _____

Created on Day _____

Draw or glue a picture of the animal above.

Specially created features _____

_____

_____

_____

Information found:
    Encyclopedia(s) _____
    On-Line _____
    Other Resources _____

Mentioned in Scripture _____

_____

Other Notes _____

_____

Animal Name _____

Vertebrate or Invertebrate _____

Warm–blooded or Cold–blooded _____

Category (Mammal, Amphibian, Reptile, Bird, Insect, Fish, Other) _____

Size (Height/Length & Weight) _____

Habitat _____

Diet _____

Enemies _____

Created on Day _____

Draw or glue a picture of the animal above.

Specially created features _____

_____

_____

_____

Information found:
    Encyclopedia(s) _____
    On-Line _____
    Other Resources _____

Mentioned in Scripture _____

_____

Other Notes _____

_____

Animal Name _____

Vertebrate or Invertebrate _____

Warm–blooded or Cold–blooded _____

Category (Mammal, Amphibian, Reptile, Bird, Insect, Fish, Other) _____

Size (Height/Length & Weight) _____

Habitat _____

Diet _____

Enemies _____

Created on Day _____

Draw or glue a picture of the animal above.

Specially created features _____

_____

_____

_____

Information found:
    Encyclopedia(s) _____
    On-Line _____
    Other Resources _____

Mentioned in Scripture _____

_____

Other Notes _____

_____

Animal Name _____

Vertebrate or Invertebrate _____

Warm–blooded or Cold–blooded _____

Category (Mammal, Amphibian, Reptile, Bird, Insect, Fish, Other) _____

Size (Height/Length & Weight) _____

Habitat _____

Diet _____

Enemies _____

Created on Day _____

Draw or glue a picture of the animal above.

Specially created features _____

_____

_____

_____

Information found:
    Encyclopedia(s) _____

    On-Line _____

    Other Resources _____

Mentioned in Scripture _____

_____

Other Notes _____

_____

Animal Name _____

Vertebrate or Invertebrate _____

Warm–blooded or Cold–blooded _____

Category (Mammal, Amphibian, Reptile, Bird, Insect, Fish, Other) _____

Size (Height/Length & Weight) _____

Habitat _____

Diet _____

Enemies _____

Created on Day _____

Draw or glue a picture of the animal above.

Specially created features _____

_____

_____

_____

Information found:
    Encyclopedia(s) _____
    On-Line _____
    Other Resources _____

Mentioned in Scripture _____

_____

Other Notes _____

_____

Animal Name _____

Vertebrate or Invertebrate _____

Warm–blooded or Cold–blooded _____

Category (Mammal, Amphibian, Reptile, Bird, Insect, Fish, Other) _____

Size (Height/Length & Weight) _____

Habitat _____

Diet _____

Enemies _____

Created on Day _____

Draw or glue a picture of the animal above.

Specially created features _____

_____

_____

_____

Information found:
    Encyclopedia(s) _____
    On-Line _____
    Other Resources _____

Mentioned in Scripture _____

_____

Other Notes _____

_____

Animal Name _____

Vertebrate or Invertebrate _____

Warm–blooded or Cold–blooded _____

Category (Mammal, Amphibian, Reptile, Bird, Insect, Fish, Other) _____

Size (Height/Length & Weight) _____

Habitat _____

Diet _____

Enemies _____

Created on Day _____

Draw or glue a picture of the animal above.

Specially created features _____

_____

_____

_____

Information found:
    Encyclopedia(s) _____

    On-Line _____

    Other Resources _____

Mentioned in Scripture _____

_____

Other Notes _____

_____

Animal Name _____

Vertebrate or Invertebrate _____

Warm–blooded or Cold–blooded _____

Category (Mammal, Amphibian, Reptile, Bird, Insect, Fish, Other) _____

Size (Height/Length & Weight) _____

Habitat _____

Diet _____

Enemies _____

Created on Day _____

Draw or glue a picture of the animal above.

Specially created features _____

_____

_____

_____

Information found:
    Encyclopedia(s) _____
    On-Line _____
    Other Resources _____

Mentioned in Scripture _____

_____

Other Notes _____

_____

Animal Name _____

Vertebrate or Invertebrate _____

Warm–blooded or Cold–blooded _____

Category (Mammal, Amphibian, Reptile, Bird, Insect, Fish, Other) _____

Size (Height/Length & Weight) _____

Habitat _____

Diet _____

Enemies _____

Created on Day _____

Draw or glue a picture of the animal above.

Specially created features _____

_____

_____

_____

Information found:
    Encyclopedia(s) _____
    On-Line _____
    Other Resources _____

Mentioned in Scripture _____

_____

Other Notes _____

_____

Animal Name _____

Vertebrate or Invertebrate _____

Warm–blooded or Cold–blooded _____

Category (Mammal, Amphibian, Reptile, Bird, Insect, Fish, Other) _____

Size (Height/Length & Weight) _____

Habitat _____

Diet _____

Enemies _____

Created on Day _____

Draw or glue a picture of the animal above.

Specially created features _____

_____

_____

_____

Information found:
    Encyclopedia(s) _____
    On-Line _____
    Other Resources _____

Mentioned in Scripture _____

_____

Other Notes _____

_____

Animal Name _____

Vertebrate or Invertebrate _____

Warm–blooded or Cold–blooded _____

Category (Mammal, Amphibian, Reptile, Bird, Insect, Fish, Other) _____

Size (Height/Length & Weight) _____

Habitat _____

Diet _____

Enemies _____

Created on Day _____

Draw or glue a picture of the animal above.

Specially created features _____

_____

_____

_____

Information found:
    Encyclopedia(s) _____
    On-Line _____
    Other Resources _____

Mentioned in Scripture _____

_____

Other Notes _____

_____

Animal Name _____

Vertebrate or Invertebrate _____

Warm–blooded or Cold–blooded _____

Category (Mammal, Amphibian, Reptile, Bird, Insect, Fish, Other) _____

Size (Height/Length & Weight) _____

Habitat _____

Diet _____

Enemies _____

Created on Day _____

Draw or glue a picture of the animal above.

Specially created features _____

_____

_____

_____

Information found:
    Encyclopedia(s) _____
    On-Line _____
    Other Resources _____

Mentioned in Scripture _____

_____

Other Notes _____

_____

Animal Name _____

Vertebrate or Invertebrate _____

Warm–blooded or Cold–blooded _____

Category (Mammal, Amphibian, Reptile, Bird, Insect, Fish, Other) _____

Size (Height/Length & Weight) _____

Habitat _____

Diet _____

Enemies _____

Created on Day _____

Draw or glue a picture of the animal above.

Specially created features _____

_____

_____

_____

Information found:
    Encyclopedia(s) _____
    On-Line _____
    Other Resources _____

Mentioned in Scripture _____

_____

Other Notes _____

_____

Animal Name _____

Vertebrate or Invertebrate _____

Warm–blooded or Cold–blooded _____

Category (Mammal, Amphibian, Reptile, Bird, Insect, Fish, Other) _____

Size (Height/Length & Weight) _____

Habitat _____

Diet _____

Enemies _____

Created on Day _____

Draw or glue a picture of the animal above.

Specially created features _____

_____

_____

_____

Information found:

    Encyclopedia(s) _____

    On-Line _____

    Other Resources _____

Mentioned in Scripture _____

_____

Other Notes _____

_____

Animal Name _____

Vertebrate or Invertebrate _____

Warm–blooded or Cold–blooded _____

Category (Mammal, Amphibian, Reptile, Bird, Insect, Fish, Other) _____

Size (Height/Length & Weight) _____

Habitat _____

Diet _____

Enemies _____

Created on Day _____

Draw or glue a picture of the animal above.

Specially created features _____

_____

_____

_____

Information found:
    Encyclopedia(s) _____
    On-Line _____
    Other Resources _____

Mentioned in Scripture _____

_____

Other Notes _____

_____

Animal Name _____

Vertebrate or Invertebrate _____

Warm–blooded or Cold–blooded _____

Category (Mammal, Amphibian, Reptile, Bird, Insect, Fish, Other) _____

Size (Height/Length & Weight) _____

Habitat _____

Diet _____

Enemies _____

Created on Day _____

Draw or glue a picture of the animal above.

Specially created features _____

_____

_____

_____

Information found:
   Encyclopedia(s) _____
   On-Line _____
   Other Resources _____

Mentioned in Scripture _____

_____

Other Notes _____

_____

Animal Name _____

Vertebrate or Invertebrate _____

Warm–blooded or Cold–blooded _____

Category (Mammal, Amphibian, Reptile, Bird, Insect, Fish, Other) _____

Size (Height/Length & Weight) _____

Habitat _____

Diet _____

Enemies _____

Created on Day _____

Draw or glue a picture of the animal above.

Specially created features _____

_____

_____

_____

Information found:
    Encyclopedia(s) _____
    On-Line _____
    Other Resources _____

Mentioned in Scripture _____

_____

Other Notes _____

_____

Animal Name _____

Vertebrate or Invertebrate _____

Warm–blooded or Cold–blooded _____

Category (Mammal, Amphibian, Reptile, Bird, Insect, Fish, Other) _____

Size (Height/Length & Weight) _____

Habitat _____

Diet _____

Enemies _____

Created on Day _____

Draw or glue a picture of the animal above.

Specially created features _____

_____

_____

_____

Information found:
    Encyclopedia(s) _____
    On-Line _____
    Other Resources _____

Mentioned in Scripture _____

_____

Other Notes _____

_____

Animal Name _____

Vertebrate or Invertebrate _____

Warm–blooded or Cold–blooded _____

Category (Mammal, Amphibian, Reptile, Bird, Insect, Fish, Other) _____

Size (Height/Length & Weight) _____

Habitat _____

Diet _____

Enemies _____

Created on Day _____

Draw or glue a picture of the animal above.

Specially created features _____

_____

_____

_____

Information found:
    Encyclopedia(s) _____
    On-Line _____
    Other Resources _____

Mentioned in Scripture _____

_____

Other Notes _____

_____

Animal Name _____

Vertebrate or Invertebrate _____

Warm–blooded or Cold–blooded _____

Category (Mammal, Amphibian, Reptile, Bird, Insect, Fish, Other) _____

Size (Height/Length & Weight) _____

Habitat _____

Diet _____

Enemies _____

Created on Day _____

Draw or glue a picture of the animal above.

Specially created features _____

_____

_____

_____

Information found:
    Encyclopedia(s) _____
    On-Line _____
    Other Resources _____

Mentioned in Scripture _____

_____

Other Notes _____

_____

Animal Name _____

Vertebrate or Invertebrate _____

Warm–blooded or Cold–blooded _____

Category (Mammal, Amphibian, Reptile, Bird, Insect, Fish, Other) _____

Size (Height/Length & Weight) _____

Habitat _____

Diet _____

Enemies _____

Created on Day _____

Draw or glue a picture of the animal above.

Specially created features _____

_____

_____

_____

Information found:
    Encyclopedia(s) _____
    On-Line _____
    Other Resources _____

Mentioned in Scripture _____

_____

Other Notes _____

_____

Animal Name _____

Vertebrate or Invertebrate _____

Warm–blooded or Cold–blooded _____

Category (Mammal, Amphibian, Reptile, Bird, Insect, Fish, Other) _____

Size (Height/Length & Weight) _____

Habitat _____

Diet _____

Enemies _____

Created on Day _____

Draw or glue a picture of the animal above.

Specially created features _____

_____

_____

_____

Information found:
    Encyclopedia(s) _____
    On-Line _____
    Other Resources _____

Mentioned in Scripture _____

_____

Other Notes _____

_____

Animal Name _____

Vertebrate or Invertebrate _____

Warm–blooded or Cold–blooded _____

Category (Mammal, Amphibian, Reptile, Bird, Insect, Fish, Other) _____

Size (Height/Length & Weight) _____

Habitat _____

Diet _____

Enemies _____

Created on Day _____

Draw or glue a picture of the animal above.

Specially created features _____

_____

_____

_____

Information found:
    Encyclopedia(s) _____
    On-Line _____
    Other Resources _____

Mentioned in Scripture _____

_____

Other Notes _____

_____

Animal Name _____

Vertebrate or Invertebrate _____

Warm–blooded or Cold–blooded _____

Category (Mammal, Amphibian, Reptile, Bird, Insect, Fish, Other) _____

Size (Height/Length & Weight) _____

Habitat _____

Diet _____

Enemies _____

Created on Day _____

Draw or glue a picture of the animal above.

Specially created features _____

_____

_____

_____

Information found:
　　Encyclopedia(s) _____
　　On-Line _____
　　Other Resources _____

Mentioned in Scripture _____

_____

Other Notes _____

_____

Animal Name _____

Vertebrate or Invertebrate _____

Warm–blooded or Cold–blooded _____

Category (Mammal, Amphibian, Reptile, Bird, Insect, Fish, Other) _____

Size (Height/Length & Weight) _____

Habitat _____

Diet _____

Enemies _____

Created on Day _____

Draw or glue a picture of the animal above.

Specially created features _____

_____

_____

_____

Information found:
    Encyclopedia(s) _____
    On-Line _____
    Other Resources _____

Mentioned in Scripture _____

_____

Other Notes _____

Animal Name _____

Vertebrate or Invertebrate _____

Warm–blooded or Cold–blooded _____

Category (Mammal, Amphibian, Reptile, Bird, Insect, Fish, Other) _____

Size (Height/Length & Weight) _____

Habitat _____

Diet _____

Enemies _____

Created on Day _____

Draw or glue a picture of the animal above.

Specially created features _____

_____

_____

_____

Information found:
    Encyclopedia(s) _____
    On-Line _____
    Other Resources _____

Mentioned in Scripture _____

_____

Other Notes _____

----

www.ingramcontent.com/pod-product-compliance
Lightning Source LLC
Chambersburg PA
CBHW081547170526
45166CB00009B/2607